USING GRAPHIC ORGANIZERS TO STUDY THE LIVING ENVIRONMENT™

Looking at How Species Compete Within Environments with Graphic Organizers

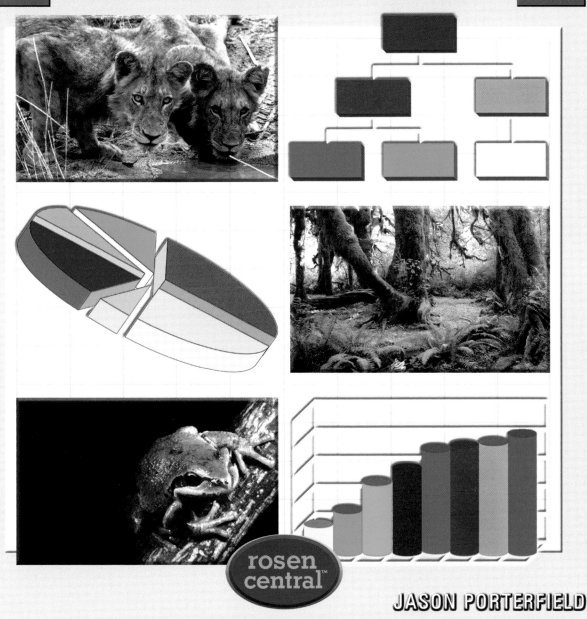

JASON PORTERFIELD

The Rosen Publishing Group, Inc., New York

Published in 2006 by The Rosen Publishing Group, Inc.
29 East 21st Street, New York, NY 10010

Copyright © 2006 by The Rosen Publishing Group, Inc.

First Edition

All rights reserved. No part of this book may be reproduced in any form without permission in writing from the publisher, except by a reviewer.

Library of Congress Cataloging-in-Publication Data

Porterfield, Jason.
Looking at how species compete within environments with graphic organizers/Jason Porterfield.
   p. cm.—(Using graphic organizers to study the living environment)
ISBN 1-4042-0613-2 (library binding)
1. Competition (Biology)—Juvenile literature. 2. Competition (Biology)—Study and teaching (Secondary)—Graphic methods. 3. Graphic organizers.
I. Title. II. Series.
QH546.3.P68 2006
577.8'3—dc22

2005022292

Manufactured in the United States of America

On the cover: A bar chart *(top right)*, a pie chart *(center)*, and a flow chart *(bottom left)*.

Contents

Introduction 4

Chapter One Species Interactions 8

Chapter Two Intraspecific Competition 15

Chapter Three Competition Between Species 22

Chapter Four Invasive Species 29

Chapter Five The Fiercest Competitor 36

Glossary 43

For More Information 44

For Further Reading 45

Bibliography 45

Index 47

Introduction

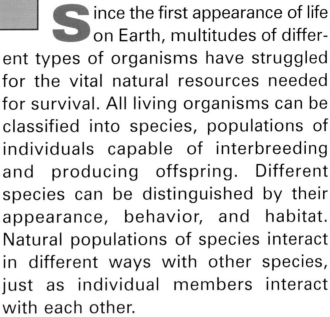

Since the first appearance of life on Earth, multitudes of different types of organisms have struggled for the vital natural resources needed for survival. All living organisms can be classified into species, populations of individuals capable of interbreeding and producing offspring. Different species can be distinguished by their appearance, behavior, and habitat. Natural populations of species interact in different ways with other species, just as individual members interact with each other.

One such interaction is species competition, which occurs throughout the natural world. Each autumn, squirrels compete against each other for nuts as they scurry through the forest to gather enough food to last until spring. A peahen selects a peacock with the largest display of tail feathers as her mate. Two similar species of lizard eat the same food, the supply of which decreases until one species is forced to leave the area.

All forms of life compete for resources in their ecosystems. Pictured here are a forest *(left)*, a large lizard *(top right)*, a deer *(middle right)*, and a bald eagle catching a fish *(bottom right)*.

Concept Web: The Five Kingdoms of Life

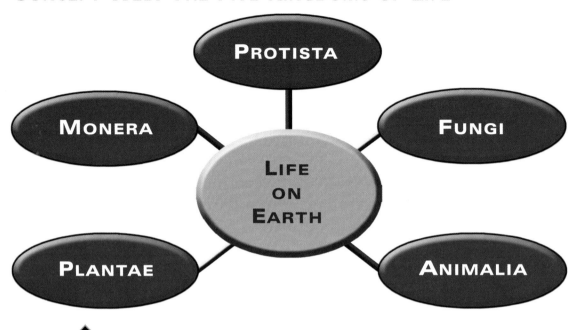

Concept webs are graphic organizers that show the relationship between terms. The subject of the web goes in the middle and is surrounded by related facts. This concept web shows the five major divisions, called kingdoms, of life on Earth.

In biology, "competition" refers to interactions between members of the same species or different species vying for the resources they need to survive. Competition takes place between all types of organisms, from two wildflowers in a field to predators, such as lions and hyenas. Individuals compete for food, water, light, social status, and mates. In intraspecific competition, individuals compete within their own species for resources and mates. Interspecific competition refers to competition between members of different species for resources. Competition can be fierce for many species, and populations often decrease through competition. The stronger individuals survive and pass on their strengths to future generations.

Species competition plays a key role in the survival of many ecosystems. It helps maintain balanced populations among the species living within the same community. Many ecosystems are delicate, and a small change can throw the entire system off balance. Invasive species that migrate from other ecosystems can displace native organisms, disrupting the interactions of other species and throwing off the competitive balance.

Human beings pose another major threat to the balance of ecosystems. Through technological innovations and population growth, humans have in many ways moved beyond directly competing with other organisms for resources; instead they simply take what they need from the environment. Many human behaviors have proven destructive to plant and animal habitats. The rapidly growing human population and humanity's need for space threaten to force many species from their ecosystems. Humans themselves also inadvertently destroy entire ecosystems through pollution and poor resource management. Today, many species are extinct or endangered as a result of human activity.

Competition can be explained using graphic organizers, tools used to present information in a way that is easy to understand. Graphic organizers come in many forms, including graphs, maps, tables, and charts. They are useful tools for understanding the effects of competition on a species and its environment.

Chapter One
Species Interactions

In order to survive, all organisms on the planet must coexist with the other life-forms in their natural environments. Some interactions between two species are harmonious, benefiting both species. Other interactions, such as the relationship between a predator and its prey, benefit one species and harm the other. The study of interactions between species is a branch of the scientific discipline known as population biology, or population ecology. Population biology also deals with dynamics within species and examines how populations interact with their ecosystems.

An ecosystem is the natural environment within which an organism lives and interacts with other organisms. Every ecosystem is unique, shaped by factors such as climate and altitude. Any given place on Earth might contain several ecosystems. A large ecosystem, such as a forest, might include many smaller and vastly different ecosystems.

Each ecosystem consists of two parts: the habitat and the community. The habitat includes the ecosystem's physical surroundings or components. The community consists of all life-forms within the ecosystem, including animals, plants, bacteria, and other organisms. Many different species may live together and interact with each other in an ecosystem. A group of individuals of the same species living in the same ecosystem is called a population. The populations of different ecosystems are often isolated from each other to some extent.

Each species population depends on the resources available within its ecosystem. In order to survive and grow, each

Tree Chart: Ecosystems

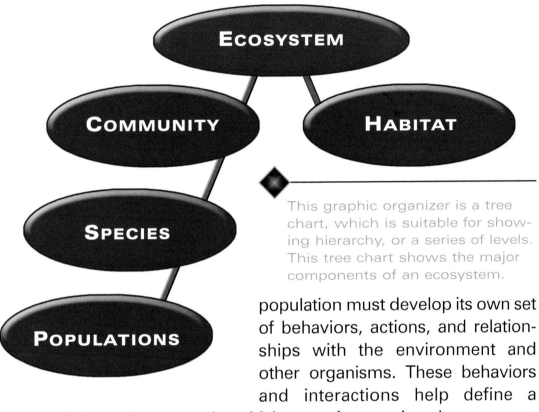

This graphic organizer is a tree chart, which is suitable for showing hierarchy, or a series of levels. This tree chart shows the major components of an ecosystem.

population must develop its own set of behaviors, actions, and relationships with the environment and other organisms. These behaviors and interactions help define a species' niche, the means by which a species acquires the resources necessary for survival and reproduction. For each species, there exists both a potential and a realized niche in the ecosystem. The potential niche, sometimes called the fundamental niche, is the behavior of the population if its growth were not limited by constraining factors. However, the growth of each species within an ecosystem is curbed by factors such as competition and limited resources. These limiting factors result in the development of a realized niche, which can shift as the species responds to a variety of environmental changes.

Types of Species Interactions

A population's niche is largely defined by the way it interacts with other species. There are many ways in which species may interact within a given ecosystem. For example, a squirrel and a bird might

Clown fishes swim by a sea anemone. Although sea anemones look like plants, they are actually flesh-eating animals.

nest in the same tree without ever coming into contact. A raccoon may live by a riverbank, where it eats fish that are too slow to elude capture. All possible types of interactions can be broken down into six main categories: neutrality, commensalism, mutualism, parasitism, predation, and competition. Each interaction has a different effect on population growth.

A neutral relationship is one in which two species may not interact directly. Nevertheless, the actions of each do affect the other. Foxes and meadow grasses seem to have little apparent connection, but they are linked through interactions with other species. When foxes eat rabbits that feed on the meadow grasses, they are inadvertently helping the grasses to grow taller. The remaining rabbits then grow fatter on the excess grasses, providing the foxes with more substantial meals.

Commensalism is an interaction that benefits one species a great deal but has little or no effect on the other species. For example, birds might build a nest within the dense branches of a shrub. The shrub gives them shelter, protects them from enemies, and provides a safe place to raise their young. The shrub gets nothing in return, but it is not harmed by the process. The birds are free to move on when they choose. The shrub will be no better and no worse off than before the birds nested.

Mutualism is the type of species interaction in which both parties benefit. Mutualistic relationships can be either short term or long term. Interactions between plants and pollinating insects are a type of short-term mutualism. The insect benefits from feeding on

SPECIES INTERACTIONS

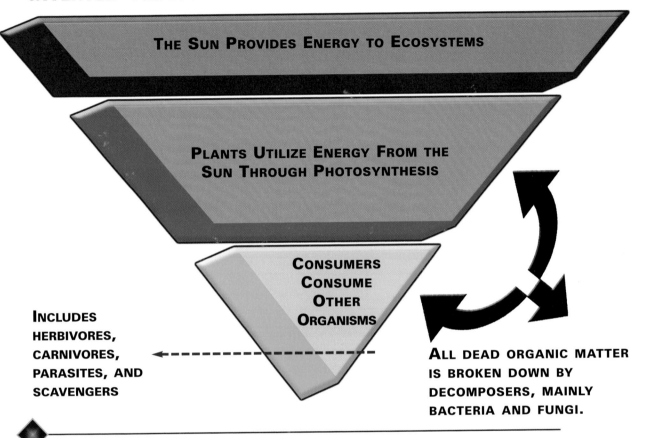

An inverted triangle is a great way to illustrate relationships. In this triangle, the sun, which is the greatest source of energy on Earth, occupies the top and widest part of the triangle. It is followed by plants, and then all other organisms.

the plant's nectar, and the insect's actions ensure that the plant population will continue to grow by gathering the plant's pollen and spreading it to more plants. An example of long-term mutualism is the relationship between a sea anemone and a clown fish. The anemone provides the clown fish with protection and a source of food, while the clown fish helps the anemone by keeping it clean.

An obligatory relationship is a specific type of long-term mutualism in which one species cannot grow without helping another, such as that between the yucca plant and the yucca moth.

Table: Outcomes of Interactions Between Two Different Species

Type of Interaction	Direct Outcome for Species 1	Direct Outcome for Species 2
Neutrality	None	None
Commensalism	Positive	None
Mutualism	Positive	Positive
Parasitism	Positive	Negative
Predation	Positive	Negative
Competition	Negative	Negative

Tables are used to group and list information so that similarities and differences between two or more things can be observed. This table shows the effects (positive, negative, or none) of the six types of interactions between species.

Yucca plants are pollinated only by yucca moths. Yucca moth larvae can grow only in yucca plants and eat only yucca seeds. Therefore, the survival of both species is dependent on their mutualistic relationship.

Parasitism and predation are both interactions that benefit one species while directly harming the other. In parasitism, one organism, the parasite, lives with another organism, the host. The parasite derives its nourishment from its living host. Parasites generally do not kill their hosts, but they can cause harm by lowering reproductive success, straining the host's immune system, or causing changes in the host's behavior. Parasites such as tapeworms usually do not directly cause a host's death and would be at a great disadvantage if the host died. Parasites can, however, indirectly kill the host by weakening its immune system to the point that an otherwise minor infection becomes deadly.

Species Interactions

Predation is the consumption of one life-form by another. Predators feed on other organisms, their prey. Unlike mutualism or parasitism, the host and prey do not live together. The success of predators depends heavily on the size of the population and the rate of reproduction of both predator and prey. The more prey living in an ecosystem, the larger the number of predators that can survive within the same ecosystem. Likewise, as the prey population decreases, the competition within the predator species causes a reduction in the predator population. When the population of prey rises again, the population of predators also rises.

Competition is a form of species interaction in which organisms must compete for resources within an ecosystem. Organisms of all descriptions compete with each other. Predators compete for prey, plants on a forest floor compete with each other for sunlight and water, and even large trees must compete with each other for space to grow. There are two main kinds of competition: intraspecific competition and

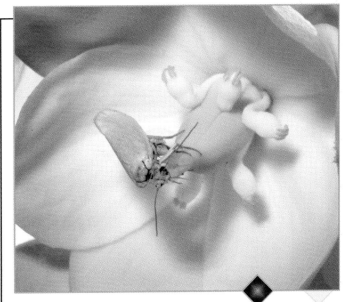

A yucca moth pollinates a yucca flower. The two species share an obligatory relationship.

A lioness chases a kudu through water in Etosha National Park in Nambia. This type of interaction is an obvious example of predation.

Looking at How Species Compete Within Environments with Graphic Organizers

Line Graph: Population Fluctuations of Predator and Prey Species

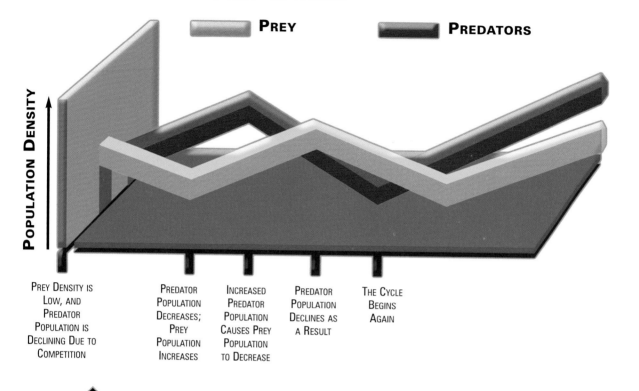

Line graphs are useful for illustrating trends. This line graph gives a general view of fluctuations in populations of predator and prey as they interact in an ecosystem.

interspecific competition. In intraspecific competition, organisms belonging to the same species compete for the same resources. Interspecific competition refers to competition between different species.

Chapter Two
Intraspecific Competition

A hardy plant crowds out a weaker neighbor, limiting its access to sunlight. A young male baboon is killed by a dominant male before it can ever mate with a female. These are two very different examples of intraspecific competition, in which individuals belonging to the same species vie for control of or access to essential resources. These resources can include food, water, sunlight, space, or mates. Intraspecific competition can become fierce, particularly in densely populated areas with limited resources.

There are two main kinds of intraspecific competition: adapted competition and unadapted competition. Adapted competition is generally based on a social structure within the species population and involves dominance issues, such as where individual organisms sleep or which individuals get to breed. Unadapted competition is unorganized. It occurs from accidental encounters between members of the same species as they seek to use the same resources, such as food, water, space, or light. An individual

Lions live in prides that are usually dominated by a single male. The dominant male protects the pride, which includes young males that are expelled just before they reach maturity. After several years, the dominant male is displaced by a stronger competitor.

organism's success in unadapted competition often affects its success in adapted competition, since the stronger competitors tend to rise in the social order. Thus, both forms of competition can have wide-ranging consequences, affecting the population, other species, and even the future development of the species.

UNADAPTED COMPETITION

Within every ecosystem, there is a limited amount of resources available for every niche. In unadapted competition, members of the same species must compete for these resources in order to survive and grow as individual organisms. This type of competition is

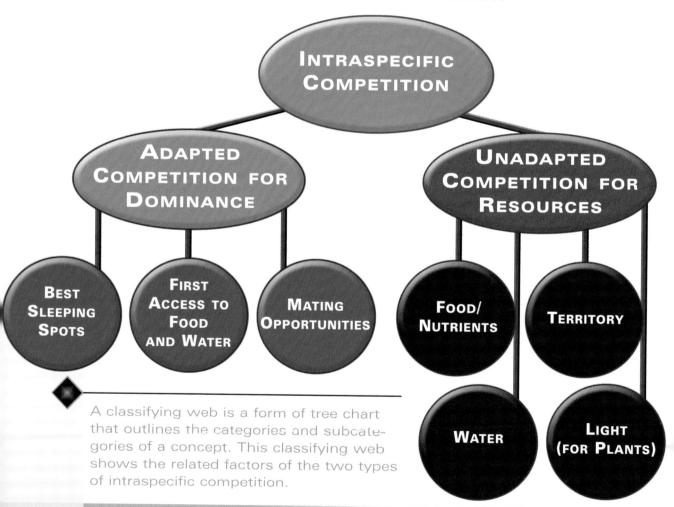

CLASSIFYING WEB: INTRASPECIFIC COMPETITION

A classifying web is a form of tree chart that outlines the categories and subcategories of a concept. This classifying web shows the related factors of the two types of intraspecific competition.

sometimes called scramble competition as all individuals strive for their own survival.

In unadapted competition, all individuals have access to resources. However, some are more capable of exploiting them than are others. The individual most capable of adapting to a given situation is most likely to succeed. If two foxes in the same ecosystem must compete for the same prey, the faster and stronger fox will likely catch more prey than the slower, weaker fox. Likewise, if the two foxes end up fighting over a rabbit, the stronger of the two will likely win. This type of interaction often reduces the supply of shared, limited resources available to weaker members of the species population. In other cases, one individual or group might assert its dominance over the rest of the population by controlling access to the resource, regardless of its abundance or scarcity.

In a densely populated ecosystem, several factors regulate the consumption of resources and the growth of the species population. When an ecosystem becomes crowded and resources grow strained, the lifespan of individuals decreases and the birthrate drops. The population of competing individuals either levels off or drops, and the resources once again become abundant. The strongest individuals and the best competitors are usually most likely to survive these fluctuations.

This pattern of population growth is particularly evident when applied to a species' food supply. For example, in seasons in which oak trees produce fewer acorns than usual, squirrels have to work harder to collect the available nuts. They react to the shortage by bearing fewer young. When oak trees produce an abundance of acorns and the squirrels don't have as much competition, they have more young the following spring. If the increased population creates a strain on the acorn supply the next fall, then fewer squirrels will survive through the winter. The survivors will be the ones who outcompeted the weaker members of the population by gathering more nuts.

Adapted Competition

Adapted competition is intraspecific competition based more on social advancement within populations than the survival of an individual. The individuals most capable of outcompeting others for

Cause-and-Effect Flow Chart: Acorn Production and a Squirrel Population

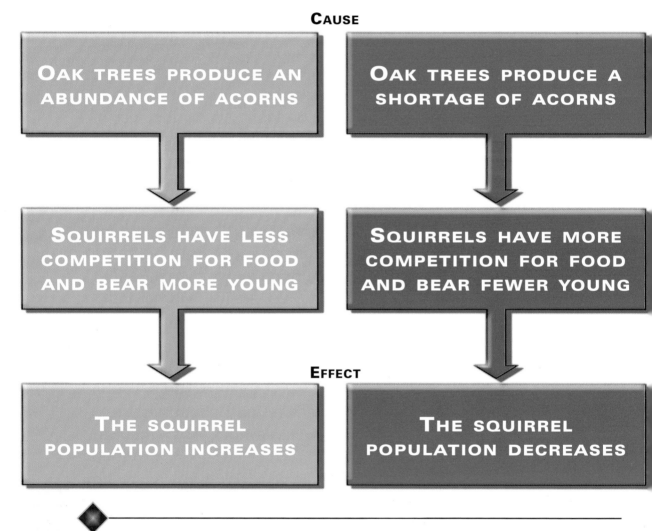

This graphic organizer is a cause-and-effect flow chart that shows the effect of acorn production on a squirrel population. Flow charts are useful in illustrating a process or the sequence of outcomes that results from an event.

food and territory are the ones most likely to rise to the top of a social hierarchy. Adapted competition usually determines which individuals within a population will mate: the stronger and best-adapted candidates typically win. Because they are the fittest competitors, their genes have a greater likelihood of ensuring the continuation of the species in future generations.

In a given population, there are typically separate dominance hierarchies for males and females, as well as both a dominant male and a dominant female. Competition for dominance takes place only between members of the same sex, though the outcome of these struggles dictates who mates with whom, therefore influencing the future of the population.

Individuals who become the dominant members of a population frequently become the first to utilize resources, such as food or water, as they become available. They sleep in the safest places. They often can control how or when other members of the population use resources.

The struggle for social dominance among many populations is seldom easy, especially among predators. In many populations, questions of dominance are resolved by fighting. Younger individuals seeking dominance in social groups such as lion prides will challenge the dominant individual to fight. If the dominant lion wins such a fight, it maintains its place at the top of the social hierarchy. The younger lion must either accept its lower status or leave the pride and join another group in which it may have a better chance of becoming dominant. If the younger lion wins, then the older lion must rechallenge for dominance, accept its lower status, or move on.

A dominant animal is usually the one most frequently selected for mating by members of the opposite sex. Competition for mates often induces extraordinary behavior in individuals, usually males, as they strive for dominance. The results of these competitions usually decide which individuals are best adapted to further the species. In the case of birds, many individuals use dances or display their feathers to attract a mate.

Chart: The Handicap Principle and Dominance

Species	Handicap	Indication of Dominance
Peacocks	Dragging tail feathers.	Animals with heavier feathers still capable of fleeing predators are stronger and more desirable.
Elk	Larger antlers are heavier than smaller antlers.	Heavier antlers require more strength to bear and are more useful in fighting, making males with large antlers more attractive mates.
White Pelicans	Develop fleshy bumps between their eyes during breeding season.	The bumps hinder the pelican's ability to see and catch prey. Pelicans that continue to hunt well are better equipped to feed a family and, therefore, make better mates.
Frogs	Nocturnal mating calls attract predators.	Frogs capable of avoiding predators despite revealing their location are chosen as better mates.

This three-column chart, or table, gives four examples of the role the handicap principle plays in social dominance.

Another factor that comes into play in the selection of a mate is its ability to either fight off or outrun predators. An individual capable of fending off or fleeing from other animals is an obvious choice for a mate, as its offspring would have a good chance of inheriting these qualities.

In some species, this form of selection, referred to as the handicap principle, is carried to extremes. Peacocks, for example, possess a spectacular plume of tail feathers, which they display to prospective mates. However, these feathers drag on the ground, when not being displayed and hinder the peacock's ability to escape from predators by slowing it down. Logically, a peahen selecting a mate should choose a peacock with smaller tail feathers because these males have a better chance of escaping. However, this is not the case. Instead, they choose the males with the larger feather displays, although they present a disadvantage to the fleeing male.

A peacock displays its spectacular tail, most likely to attract a peahen. The peahen's tail is not nearly as impressive, but she can fan it out to protect her chicks.

The idea behind the handicap principle is that the males with the larger feather displays who can still escape from enemies and survive to breeding age are the strongest members of the population. Their "handicap" forces them to grow stronger than males with smaller displays in order to survive. Therefore, the males with the larger displays are those most likely to ensure the continuation of the species.

Chapter Three
Competition Between Species

Many ecosystems are capable of supporting a wide variety of life. When there are significant differences in the needs of individual species, a greater number of different species populations can successfully coexist. When niches overlap, organisms are forced to compete not just with other members of their own population, but with other populations as well.

COMPETITIVE EXCLUSION PRINCIPLE

Interspecific competition occurs when two different species living within the same ecosystem must compete with each other for the same resources. Typically, the more two species differ from each other, the likelier they are to coexist within the same environment. They either need different resources, or they utilize resources in such a different way that they don't have to interact with each other. However, two species that fill similar niches and use resources the same way are less likely to live in the same ecosystem.

When a new organism is introduced into an ecosystem with the same needs as an organism already in place, the two populations will not be able to coexist indefinitely. The idea that two species requiring identical resources cannot coexist indefinitely is referred to as the competitive exclusion principle. One species will eventually outcompete the other for food and territory, thereby forcing the other species to move into another territory. Two species of rabbit that eat exactly the same kind of grass would not be able to survive in the same ecosystem for

Venn Diagram: Diets of Two Competing Species

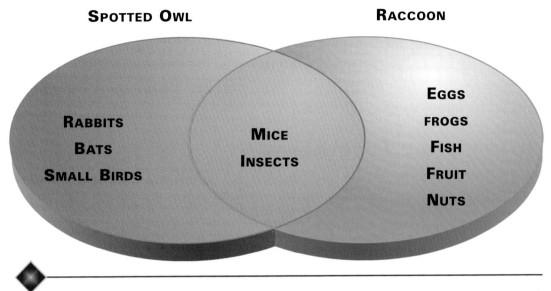

Venn diagrams are useful in examining similarities and differences. Similarities fall in the overlapping parts of the circles. This Venn diagram shows the diets of the spotted owl and the raccoon.

very long. One species would have to either adapt to the situation and eat other grasses, move into another ecosystem, or die out.

COMPETITIVE RELEASE

When two species in an ecosystem find themselves competing for the same resources, one of the species is generally not capable of outcompeting the other. To compensate, that species will instead begin breeding more frequently. The rapid population growth that follows is called competitive release. Competitive release usually occurs when the superior competitor's own population is low. This gives the inferior species an opportunity to dominate the ecosystem's resources and outcompete its rival through sheer numbers alone. This type of competitive release occurs most frequently when a species is introduced from one ecosystem to another.

Another form of competitive release occurs after a competing species has been eliminated from an ecosystem. When two species compete for the same resources, the population growth of both species usually shrinks. More individuals die off earlier and the birth rate falls because the stressed ecosystem cannot support unrestrained growth of either species. When competitive exclusion forces one species out, more resources become available to the remaining species. As a result, that species' population increases rapidly to fill its expanded niche.

COMPETITIVE DISPLACEMENT

Competitive displacement is the removal of an established species from an ecosystem as a result of either direct or indirect competition from another species. There are three ways in which competitive displacement can occur. A species that comes to dominate a niche within an ecosystem may drive the losing species out completely or even kill it off. Alternately, one species may differ significantly from its competitor in its growth and use of resources. In this scenario, the population of one competing species may simply grow too slowly to keep up with another competitor, particularly if the faster-growing population is experiencing competitive release. The third form of competitive displacement occurs through random chance. One population may simply decide to shift to another ecosystem, leaving its competitor to exploit any resources in the original ecosystem.

If a species displaced by competition is to survive, it often has to utilize resources within a less advantageous habitat. The species may adapt to eat slightly different foods or nest in different ways. Plant species can also adapt to habitats that are less desirable.

Scientists studying the silver maple tree in eastern North America have found that in the wild, it grows mostly in forested wetlands or swamps. However, the silver maple grows faster and larger and produces more fruits in well-drained forests. For this reason, many cities and towns choose to plant silver maples along streets and in parks. Although it grows better on well-drained land, it doesn't naturally occur in these places because it cannot successfully compete with other tree species, such as the sugar maple and

Table: Causes of Competitive Displacement

Dominance of resources by one species	One competing species gains greater access to necessary resources and displaces the other.	Reef corals poison and grow over other reef corals to control an area.
Gradual replacement	One species grows more slowly and is gradually displaced by the other.	Sugar maples grow faster than silver maples, displacing them to swamps.
Displacement by chance	One species decides to leave the ecosystem for another.	Bears wander to another forest in search of their favorite food.

◆───────────────────────────────

This table provides a snapshot of three causes of competitive displacement, giving an example for each. Organizing information in this way can make it easier to understand and remember.

the red oak, found in these places. Thus, the silver maple has been displaced to occupy a new niche for itself in swamps.

Over a long period of time, competitive displacement can result in evolutionary changes in a species. This occurs when displaced species are forced to evolve new behaviors or new physical characteristics in order to survive in suboptimal conditions. Eventually, the displaced species may develop into an entirely new species.

When the naturalist Charles Darwin was developing his theory on evolution, he based much of his research on a bird called the Galapagos finch, of which Darwin recorded fourteen different species. According to Darwin, the different species of Galapagos finch are all descended from a single species of finch, which had left

Charles Darwin outlined his theory of evolution in 1859 in a book called *On the Origin of Species*. He is often referred to as the father of modern biology.

South America and colonized the Galapagos Islands many thousands of years ago. The finch population then expanded rapidly, straining the ecosystem. Individual birds that were less successful competitors were forced to survive in less optimal habitats. As a result, the Galapagos finches developed unique physical characteristics for life in these places.

The original ancestor finches probably had a short, cone-shaped bill, adapted for crushing seeds. As the species dispersed throughout the different habitats of the Galapagos Islands, the bird populations in these marginal environments adapted to eating different things, such as fruit or insects. Over time, these populations evolved beaks specifically shaped for the new types of food they had to eat because of competitive displacement. Today, the fourteen species of finch found on the Galapagos Islands are easily distinguished from each other by the size and shape of their bills.

RESOURCE PARTITIONING

Even when resources are scarce and two species in the same ecosystem share similar needs, it is sometimes possible for them to coexist. Similar species do this by subdividing resources among

Like most webs, a cluster web shows the relationship between things or concepts. This cluster web gives the scientific names and several characteristics of the fourteen species of finch that evolved from the original ancestor finch.

Cluster Web: Types and Characteristics of Galapagos Finches

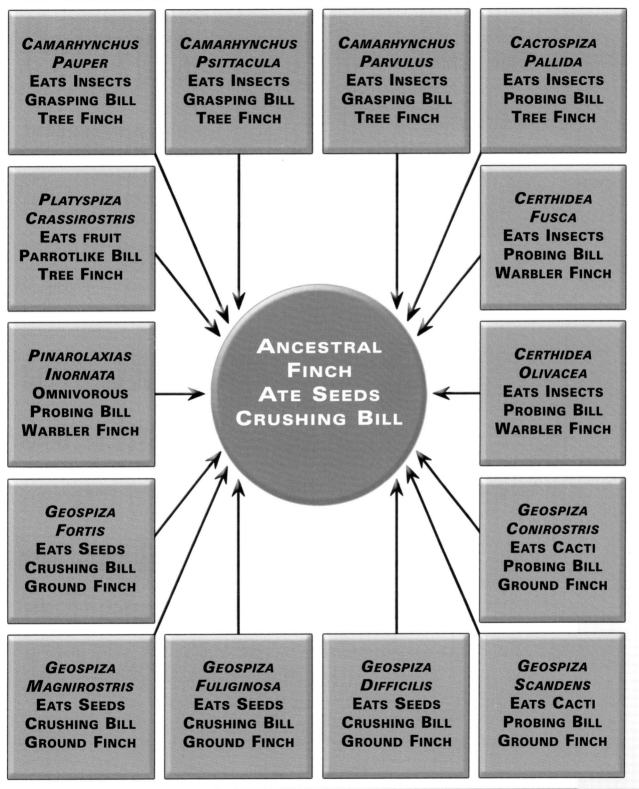

T-Chart: Anole Tree Partitioning

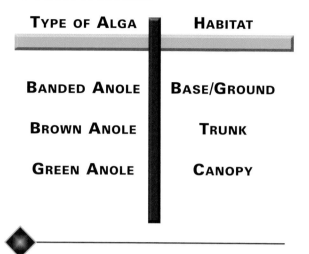

Type of Alga	Habitat
Banded Anole	Base/Ground
Brown Anole	Trunk
Green Anole	Canopy

A T-chart is a simple two-column table in the shape of the letter T. This T-chart shows the specific section of the trees on which each species of Caribbean anole feed in South American forests.

themselves, a process called resource partitioning. Two competing species may use the same resources but at different times. Other species use different parts of the same resources to reduce competition. For example, three different meadow plants—bristly foxtail, indian mallow, and smartweed—all have similar needs. Each requires sunlight, water, and dissolved minerals found in the soil. Yet each plant species is adapted to exploit different portions of the habitat at different times.

The foxtail grasses have shallow, drought-resistant roots that absorb water quickly and grow where moisture levels vary widely from day to day. Mallow plants have a long taproot designed to reach deep into the ground, where the soil stays moist later in the growing season. It grows where topsoil is moist early in the season and gradually dries as the season progresses. The smartweed has a long taproot but also has many shorter roots close to the surface and grows where the soil is continually moist. Because each plant has a different way of exploiting moisture and needs moisture at different times, they can all live together in the same ecosystem.

Similarly, three different species of Caribbean anoles (small, chameleon-like lizards) partition resources in South American forests. All three species eat insects. Individuals belonging to each species often live in the same tree, but they minimize competition for insects by partitioning the sections of the tree. Each utilizes the resources in its own niche, ignoring those of the other anole species.

Chapter Four
Invasive Species

As seen in previous chapters, nature maintains a natural balance among species populations through competition. Occasionally, species that have been successful in their own ecosystems will move from their habitat and attempt to establish themselves elsewhere. The outcome of these population shifts can vary drastically. If the species resembles another species already established in the new ecosystem, then interspecific competition will likely force one to leave or die out. If the new species outcompetes the old, then it simply takes the old species' place in the ecosystem. But if the new species moves in and begins exploiting an empty niche in the ecosystem, it can expand rapidly, using up valuable resources and throwing the ecosystem out of balance. These species are referred to as invasive species.

Geographic Dispersal

The process that brings successful species to new habitats is called geographic dispersal. Geographic dispersal occurs in three ways. A population may simply expand its territory by gradually extending its range into neighboring regions. In the second means of geographic dispersal, jump dispersal, a species is rapidly transported across great distances to a new habitat. Jump dispersal often carries individuals across inhospitable regions, such as when insects are carried across oceans in the holds of cargo ships. The third means of geographic dispersal closely resembles the first, in that a species may gradually shift to another habitat, abandoning the old one with almost imperceptible slowness.

Geographic dispersal, particularly jump dispersal, can take place with astonishing rapidity. Since Europeans first arrived in North America some 500 years ago, more than 4,500 known non-indigenous species have established themselves there. Some, such as potatoes, corn, and rice, have been introduced without any negative consequences for the environment. Others have quickly spread throughout their new ecosystem, sometimes devastating native species populations.

The water hyacinth, for example, was introduced to North America at a South American plant exposition held in New Orleans, Louisiana, during the 1880s. People struck by its lovely blue flowers took clippings from it and planted them in ponds and waterways. Fast-growing and hardy competitors, the hyacinths quickly took over many aquatic habitats, displacing many native species. Today, water hyacinths have spread throughout ponds and slow-moving rivers as far west as San Francisco, California.

Another invasive plant, kudzu, was imported from Japan in 1876. A vine, it grows best in warmer climates and in sunny spots where the soil has been recently disturbed. Between 1935 and 1953, it was widely planted on hillsides, in fields, and along roadsides to prevent erosion in the American South. However, the plant spreads very rapidly and its growth is difficult to control. Its vines rapidly take over an area, choking out other plants and making the land unusable. Today it occupies about 20,000 square miles (51,800 square kilometers) of land in the United States, mostly in the Southeast, though it has been found as far north as Pennsylvania and as far west as Oregon.

Zebra mussels were first discovered in the waters of the Great Lakes in 1988. They probably arrived on a cargo ship's hull. Since then, they have displaced numerous native shellfish populations and spread into other waterways, such as the Mississippi and Hudson rivers.

Table: Negative Effects of Introduced Species in the U.S.

Species	Origin	Form of Introduction	Outcome
Argentine Fire Ant	Argentina	Accidental (1891)	Damages assorted crops; destroys native ant populations; kills ground-nesting birds
Camphor Scale Insect	Japan	Accidental (1920s)	Damages roughly 200 species of plants in Alabama, Louisiana, and Texas
Carp	Germany	Intentional (1887)	Displaces native fish; uproots water plants, thereby reducing waterfowl populations
Chestnut Blight Fungus	Asia	Accidental (1900)	Destroyed most Eastern American chestnuts; disrupted forest ecology
Dutch Elm Disease	Europe	Accidental (1930)	Destroyed millions of elms; disrupted forest ecology
European Starling	Europe	Intentional (1890)	Competed with native songbirds; damaged crops; transmitted swine diseases
European Wild Boar	Europe	Intentional (1912)	Destroyed habitat by rooting; damaged crops
House Sparrow	England	Intentional (1853)	Damaged crops; displaced native songbirds; transmitted diseases
Japanese Beetle	Japan	Accidental (1911)	Defoliated more than 250 species of trees and plants.
Nutria	Argentina	Intentional (1940)	Altered marsh ecology; destroyed crops
Water Hyacinth	South America	Intentional (1884)	Clogged waterways, shading out other vegetation

This four-column chart shows the origin, form of introduction, and outcome for each of eleven nonnative species that were introduced into the United States between 1853 and 1940.

Unlike zebra mussels, the gypsy moth was intentionally brought into the United States from France in 1868 by scientists hoping to breed a better silkworm. The moth escaped and today can be found in forests throughout the United States. They lay their eggs in branches and under tree bark. When the young hatch, they eat the tree's leaves until it is stripped bare and then they move on to the next tree, a process that can severely damage a tree. Gypsy moths breed rapidly and are capable of outcompeting most native species. Since 1980, they have defoliated more than 1 million acres (404,686 hectares) of forest annually.

One of the world's most famous instances of a species invasion occurred in 1859, when an Australian landowner released a few wild English rabbits on his farm with the intention of hunting them. The habitat proved ideal for the rabbits, particularly since they had no natural enemies in Australia. Within six years, the rabbit population had exploded at a frightening rate, growing from only 24 rabbits to more than 40,000.

The rabbits displaced livestock and native animals, such as kangaroos. A fence 1,136 miles long (1,828 km) was built to keep them restricted to the north, but the rabbit population had expanded to the other side before the fence was complete. In 1951, a rabbit virus called myxoma was introduced through another rabbit population brought from South America. Most of the English rabbits had no natural immunity and quickly began dying off. However, populations of virus-resistant rabbits continue to grow rapidly in Australia, continuing to

Zebra mussels crowd out other aquatic life-forms in two ways: the female produces between 30,000 and 100,000 eggs per year, which enables them to multiply rapidly. They are also voracious eaters that consume most of the food for microscopic aquatic animals, setting off a chain of reaction, that starves larger fish.

cause problems for farmers and native species.

INVASIONS AND POPULATION CHANGES

Invasions by nonnative species can have a tremendous impact on species apart from the ones they feed on or outcompete. For some native species, the impact can even be positive. In 1900, the chestnut blight fungus entered the United States through nursery plants brought from Asia. The blight infected the eastern American chestnut population, and by the early 1950s the American chestnut was all but eliminated from North American forests. Prior to the blight, the American chestnut had been one of the dominant fixtures in eastern deciduous forests.

◆

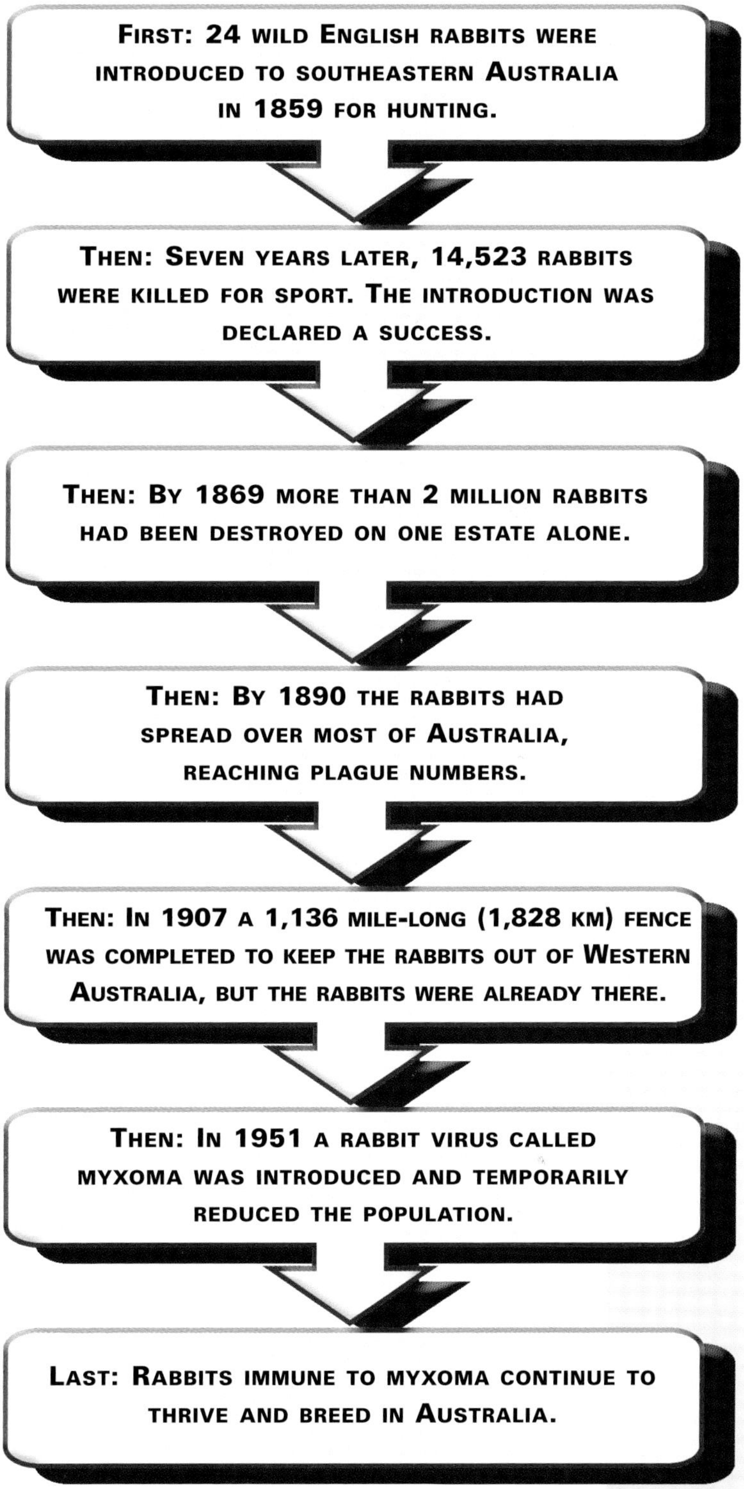

Sequence charts are useful in giving a chronological progression of cause-and-effect events. This chart shows the major developments that followed the introduction of wild English rabbits in Australia in 1859.

SEQUENCE CHART: RABBITS IN AUSTRALIA

FIRST: 24 WILD ENGLISH RABBITS WERE INTRODUCED TO SOUTHEASTERN AUSTRALIA IN 1859 FOR HUNTING.

THEN: SEVEN YEARS LATER, 14,523 RABBITS WERE KILLED FOR SPORT. THE INTRODUCTION WAS DECLARED A SUCCESS.

THEN: BY 1869 MORE THAN 2 MILLION RABBITS HAD BEEN DESTROYED ON ONE ESTATE ALONE.

THEN: BY 1890 THE RABBITS HAD SPREAD OVER MOST OF AUSTRALIA, REACHING PLAGUE NUMBERS.

THEN: IN 1907 A 1,136 MILE-LONG (1,828 KM) FENCE WAS COMPLETED TO KEEP THE RABBITS OUT OF WESTERN AUSTRALIA, BUT THE RABBITS WERE ALREADY THERE.

THEN: IN 1951 A RABBIT VIRUS CALLED MYXOMA WAS INTRODUCED AND TEMPORARILY REDUCED THE POPULATION.

LAST: RABBITS IMMUNE TO MYXOMA CONTINUE TO THRIVE AND BREED IN AUSTRALIA.

LOOKING AT HOW SPECIES COMPETE WITHIN ENVIRONMENTS WITH GRAPHIC ORGANIZERS

TIMELINE WHEEL: SPREAD OF DUTCH ELM DISEASE

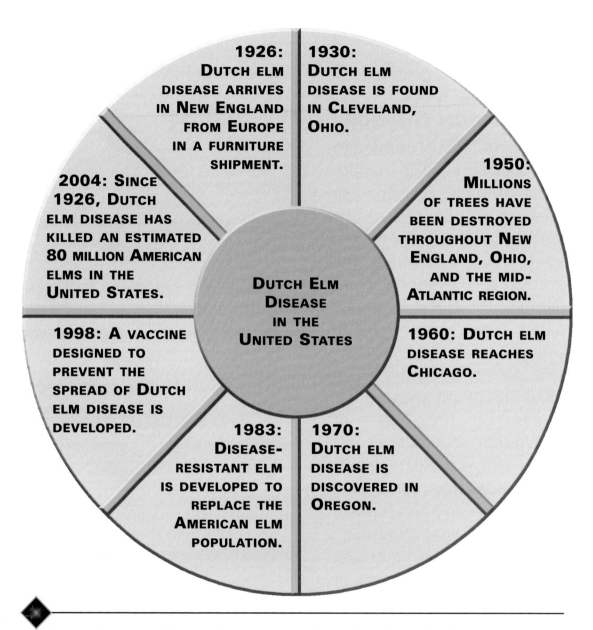

A timeline wheel is another way to show the chronological progression of events. Unlike in sequence charts, the succeeding events of a timeline wheel need not have a causal relationship.

When the tree died out, it left its niche open for other competitors. Many other tree species took advantage of the chestnut's disappearance from the forest canopy. Through competitive release, they quickly moved in to fill the chestnut's place.

Invasive Species

Possibility Tree: Possible Outcomes of Species Introduction

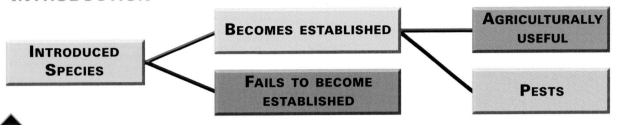

This chart is a possibility tree that shows the possible outcomes of a new species being introduced into an ecosystem.

Similarly, a fungus that causes Dutch elm disease was accidentally imported to the United States in a load of European lumber in 1930. Elm trees native to the United States had no defense against the fungus, and it quickly spread throughout the American Northeast, killing millions of elm trees and disrupting forest ecosystems. The destruction of American elm populations left an empty niche in many forests open to be filled by competitors.

Gypsy moth populations cycle between periods when they are plentiful in an area and periods when the population drops due to competitive stress. When the population is lower, predators can severely damage a gypsy moth population. A variety of insects, spiders, small mammals, and birds eat gypsy moth larvae. The presence of a gypsy moth population in an area often attracts predators, such as red-winged blackbirds. The presence of these additional predators can strain the local ecosystems, as these animals compete against those usually found in the area.

The gypsy moth was introduced into the United States in 1868. It quickly established itself in locations in the Northeast, and is now one of the most destructive pests of hardwood forests.

Chapter Five
The Fiercest Competitor

Over the billions of years that life has existed on Earth, many species have emerged and become extinct. Others successfully adapted to environmental changes and evolved into present-day species. Still others, such as crocodiles and sharks, have changed very little over hundreds of millions of years. Yet after surviving catastrophic events such as ice ages and asteroids, many species on Earth now face a new challenge from human activities and rapid population growth. The interactions between people and the environment put many species into direct competition with humans for space and other resources.

Humans and Resources

If placed in a natural environment without the support of technology, modern humans wouldn't appear very threatening. Compared to many species, modern humans are slow, awkward, and ill-equipped for hunting or defense. Our lives would likely resemble those of other social primates and be spent foraging for food and avoiding predators. However, human ancestors discovered how to make and use tools, build fires, construct shelters, and unite in organized societies against predators and competitors. Today, humans are the dominant species on Earth.

Like all mammals, humans require food, air, water, space, and sunlight. However, technology has allowed people to exploit resources in nearly any habitat. We mine coal and minerals from the earth, drill for oil, strip-mine mountaintops, and clear tracts of forest for timber and land. Native species have no chance of competing against human technology, intelligence, and sheer numbers.

In a sense, humans can be considered an invasive species that has spread into ecosystems across Earth. By entering into habitats that were once free of humans and overexploiting resources, people frequently bring about the collapse of local ecosystems and a loss of species diversity. As the human population becomes more concentrated in urban areas, the world's cities and towns expand outward into previously undeveloped land. This urban sprawl contributes to the loss of a variety of habitats.

Humans currently use about 21 percent of Earth's land surface for agriculture. Poor farming techniques contribute to the destruction of prairies and grasslands. Overgrazing by livestock destroys native grasses and compacts the soil. In a drought, the bare topsoil can just blow away as it did in sections of the American

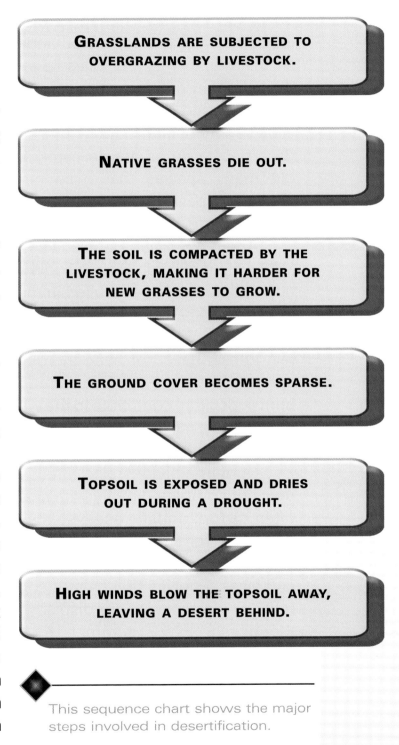

Sequence Chart: How Desertification Occurs

- Grasslands are subjected to overgrazing by livestock.
- Native grasses die out.
- The soil is compacted by the livestock, making it harder for new grasses to grow.
- The ground cover becomes sparse.
- Topsoil is exposed and dries out during a drought.
- High winds blow the topsoil away, leaving a desert behind.

This sequence chart shows the major steps involved in desertification.

Midwest during the 1930s. Since the 1930s, about 5.5 million square miles (14.2 million sq km) of grasslands worldwide have been turned into desert, and another 124,000 square miles (32,159 sq km) are converted annually.

Farms, towns, and cities have encroached on desert land with the help of irrigation systems fed by distant rivers. Though they are often depicted as lifeless places, desert ecosystems are home to a wide variety of plants and animals. Because of the scarcity of water, desert ecosystems are delicately balanced. Human activities force out many species that once thrived in certain desert regions. Displaced species move into other desert lands, increasing the populations there, straining resources, and intensifying competition.

About 4.5 billion people live on just 10 percent of Earth's land. However, population growth and the need for more agricultural products and timber have resulted in the increase of farming and logging in forested regions. Rain forests, some of the most diverse habitats in the world, contain two-thirds of all plant and animal species. Since rain forests have not been thoroughly explored by humans, many species have yet to be discovered. But practices such as slash-and-burn agriculture contribute to the wholesale destruction of the world's forests and a massive loss of species diversity. The plants and animals that lived there are displaced, many species driven to extinction.

Human activities also affect the air and water. Although we may not compete directly with species in aquatic habitats, humans often disrupt aquatic ecosystems, depriving species of resources necessary for survival. We build dams in waterways, pump water from lakes and oceans, and dump chemicals and waste in rivers and lakes. Species that cannot cope with the changes in river currents or water levels are displaced. Pollutants eventually travel to the ocean. Every summer, for example, a 7,000-square-mile (18,130 sq km) lifeless zone develops in the Gulf of Mexico. It is caused by pollutants originating from farms and factories along the Mississippi River. Airborne contaminants can also cause health problems that affect many plant and animal species.

Table: Deforestation

Region	Total Land Area	Total Forest Cover	Forest Cover	Change in Forest Cover	Rate of Change (Deforestation Rate)
	2000	2000	2000	1999–2000	1999–2000
	('000 HA)	('000 HA)	% of land	('000 HA)	(%)
Africa	2,978,394	649,866	21.8	-5,262	-0.8
Asia	3,084,746	547,793	17.8	-364	-0.1
Europe	2,259,957	1,039,251	46.0	881	0.1
North and Central America	2,136,966	549,304	25.7	-570	-0.1
Oceania & Australia	849,096	197,623	23.3	-365	-0.2
South America	1,754,741	885,618	50.5	-3,711	-0.4
Antarctica	1,400,000	0	0	0	0.0

Source: Food and Agriculture Organization of the U.N.: *The State of the World's Forests 2003*.

This table provides information about forest cover and the rate of deforestation on all continents in 2000. Europe is the only continent to have gained forest cover between 1999 and 2000.

Humans fish the oceans on a massive scale. Species that feed primarily on fish find themselves in direct competition with people. Many of these predator species are displaced as overfishing in the world's oceans depletes populations, threatening populations of both predators and prey.

Human Impact on Species Diversity

Humans effectively outcompete native species for water, space, and food, upsetting many of the world's ecosystems. Thousands of species have gone extinct or become endangered through human activity. In many cases, extinctions of species have been unintentional. In others, people have set out to eliminate given species from a certain area.

Overhunting of game has led to some of the most famous animal extinctions. The quagga, a type of zebra that once lived in southern Africa, was hunted to extinction for its beautiful coat in 1883. The moa, a large, ostrichlike bird, had been hunted to extinction in New Zealand by the time English explorer James Cook landed on the island in 1769. The passenger pigeon, once one of the most numerous game birds in North America, rapidly declined toward the end of the nineteenth century as a result of overhunting.

In other instances, humans have hunted animals that they considered dangerous or destructive. The Tasmanian wolf, a fierce predator, was viewed as a threat to livestock and its population was decimated by hunters. The last Tasmanian wolf is believed to have died in an Australian zoo in 1936. Farmers throughout eastern Europe despised and hunted the tarpan, a small species of wild horse, because it destroyed crops and led domestic horses astray. The last tarpan died in a Moscow zoo in 1875.

Many other species are endangered due to habitat loss. The mountain gorilla in Africa is threatened by humans logging rain forests. The giant panda in China could become extinct through the destruction of bamboo forests, its main source of food. South American orchid species could vanish due to logging.

Within many biological communities, there is a single species, called the keystone species, that plays a vital role in the lives of other species. It may act as a population check by consuming other species, or it may provide food or important nesting habitats for

TIMELINE: RECENT SPECIES EXTINCTIONS CAUSED BY HUMANS

1769	Moa hunted to extinction in New Zealand.
1850	The great auk hunted to extinction.
1875	The last tarpan died in a Moscow zoo.
1883	The quagga is hunted to extinction in Africa.
1911	Passenger pigeon is hunted to extinction in North America.
1936	The Tasmanian wolf becomes extinct in Australia.
1989	The golden toad is declared extinct.
2000	The red colobus monkey is declared extinct.

This graphic organizer is a timeline listing several species of animals that have become extinct due to human activity since 1769.

other organisms. When this organism is removed from the ecosystem, the rest of the community is changed dramatically.

A single keystone species can play a different role from one ecosystem to another. For example, the periwinkle snail can either increase or decrease species diversity in different settings. Those living in tide pools feed on a dominant species of alga, allowing populations of weaker competing algae to grow and increase the

LINE GRAPH: TIDE POOL DIVERSITY

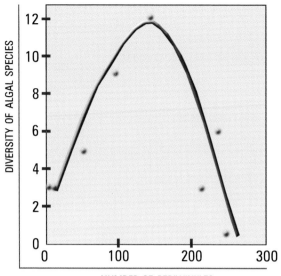

This line graph illustrates the results of a study that show that the number of algae species in tide pools is greatest when a moderate number (neither too high nor too low) of algae-eating periwinkles are present. This results in greater species diversity in a community.

community's diversity. In a slightly different ecosystem, tougher algae species are dominant on rocks that are only exposed at low tide. Periwinkles living in those ecosystems leave the tougher algae alone and eat the weaker algae species, thus reducing diversity.

When humans disrupt a keystone species population, the entire ecosystem becomes unbalanced. Gray wolves were once a keystone species in North America, regulating the populations of rabbits, deer, and other crop-eating species. Over the last 400 years they have been hunted or trapped to near extinction. Once the wolves were gone, populations of deer, rabbits, and other crop-eating species increased dramatically.

According to the World Conservation Union, 784 species have become extinct since 1500, and 1,676 others are presently endangered. Most extinctions have come about through human activity, as have the conditions that continue to lead to the endangerment of many others. Unless we take measures to preserve the diversity of wildlife, humans, nature's fiercest competitor, may find themselves much more alone on Earth.

Glossary

blight A plant disease or injury resulting in withering and death.
dispersal The spreading of organisms from one place to another.
displacement The removal of something from its natural environment.
dominance A position of superiority in a hierarchy.
ecology The study of organisms within their natural environments.
evolution The development of a biological group, such as a species, through successive generations.
extinction The end of the existence of a species.
hierarchy A classification by rank of people or things within a system.
interspecific Occurring between different species.
intraspecific Occurring within a species; involving members of one species.
niche The biological role of a species within its environment and community.
obligatory Required.
organism An individual life-form.
population A group of organisms of the same species populating a specific habitat.
species A biological group of individuals capable of interbreeding and producing offspring.
status The relative position in a hierarchy.

For More Information

National Wildlife Federation
11100 Wildlife Center Drive
Reston, VA 20190-5362
(800) 822-9919
Web site: http://www.nwf.org

Smithsonian National Museum of Natural History
Tenth Street and Constitution Avenue NW
Washington, DC 20560
(202) 633-1000
e-mail: Info@si.edu
Web site: http://www.mnh.si.edu

U.S. Fish and Wildlife Service
Endangered Species Program
4401 N. Fairfax Drive, Room 420
Arlington, VA 22203
Web site: http://endangered.fws.gov

Web Sites

Due to the changing nature of Internet links, the Rosen Publishing Group, Inc., has developed an online list of Web sites related to the subject of this book. This site is updated regularly. Please use this link to access the list:

http://www.rosenlinks.com/ugosle/scgo

For Further Reading

Gallant, Roy A. *Wonders of Biodiversity*. New York, NY: Benchmarks Books, 2003.

Gould, Stephen Jay, ed. *The Book of Life: An Illustrated History of the Evolution of Life on Earth*. New York, NY: W. W. Norton, 2001.

Mackay, Richard. *The Penguin Atlas of Endangered Species*. New York, NY: Penguin Books, 2002.

Morgan, Sally. *Ecology and Environment: The Cycles of Life*. New York, NY: Oxford University Press, 1995.

Raven, Peter H. *Biology*. New York, NY: McGraw Hill, 2001.

Bibliography

Baskin, Yvonne. *A Plague of Rats and Rubbervines: The Growing Threat of Species Invasions*. Washington, DC: Island Press, 2002.

Cohen, Joel E. *Food Webs and Niche Space*. Princeton, NJ: Princeton University Press, 1978.

Colinvaux, Paul. *Why Big Fierce Animals Are Rare: An Ecologist's Perspective*. Princeton, NJ: Princeton University Press, 1978.

Dudley, William, ed. *Biodiversity*. San Diego, CA: Greenhaven Press, 2002.

Elton, Charles S. *The Ecology of Invasions by Animals and Plants*. Chicago, IL: The University of Chicago Press, 2000.

Kormondy, Edward J. *Concepts of Ecology*. Englewood Cliffs, NJ: Prentice-Hall, 1984.

Mooney, Harold A., and Richard J. Hobbs, eds. *Invasive Species in a Changing World*. Washington, DC: Island Press, 2000.

Novacek, Michael J., ed. *The Biodiversity Crisis: Losing What Counts*. New York, NY: New Press, 2001.

Pontin, A. J. *Competition and Coexistence of Species*. Boston, MA: Pitman Advanced Publishing Program, 1982.

Slobodkin, Lawrence B. *A Citizen's Guide to Ecology*. New York, NY: Oxford University Press, 2003.

Starr, Cecie. *Biology: Concepts and Applications*. Belmont, CA: Wadsworth Publishing Company, 1997.

Van Driesche, Jason, and Roy Van Driesche. *Nature Out of Place: Biological Invasions in the Global Age*. Washington, DC: Island Press, 2000.

Zahavi, Amotz, and Avisag Zahavi. *The Handicap Principle: A Missing Piece of Darwin's Puzzle*. New York, NY: Oxford University Press, 1997.

Zimmer, Carl. *Evolution: The Triumph of an Idea*. New York, NY: HarperCollins Publishers, 2001.

Index

A
agriculture, 37, 38

C
commensalism, 10
competition, definition of, 5–6, 10, 13–14
competitive displacement, 7, 24–26, 38
competitive exclusion, 22–23, 24
competitive release, 23–24, 34
competitive stress, 35
Cook, James, 40

D
Darwin, Charles, 25–26
diversity, 38, 41–42

E
ecosystems
  disruption/destruction of, 7, 32, 35, 38, 40, 42
  parts of, 8
extinct and endangered species, 7, 40, 42
evolution of species, 25, 36

G
geographic dispersal, 29–33

H
human beings, threat of, 7, 36–42

I
interaction, types of, 8, 9–14
interbreeding, 5
interspecific competition, 6, 14, 22, 29
intraspecific competition, 6, 13, 14
  adapted, 15, 16, 18–21
  unadapted (scramble), 15–17
invasive species, 7, 29–33, 37
  and populations, 33–35

K
keystone species, 40–42

M
mate selection, 19, 20
  handicap principle of, 20–21
mutualism, 10–12, 13

N
neutrality, 10
niches, 9, 16, 24, 28, 29, 34, 35
  overlapping of, 22

P
parasitism, 10, 12, 13
population biology/ecology, 8

population growth, 7, 10, 13, 17, 23–24, 38
predation, 10, 12, 13

R
resource partitioning, 26–28

S
social dominance, 17, 19, 36

T
technology, 7, 36, 37

About the Author

Jason Porterfield is a writer and researcher who lives in Chicago, Illinois. He has written more than a dozen books on a wide range of subjects. He acquired firsthand experience with a great diversity of wildlife while growing up in rural Virginia, and as a child was a frequent visitor to the University of Virginia's Biological Station at Mountain Lake.

Photo Credits

Cover, p. 1 © Royalty-Free/Nova Development Corporation; pp. 4–5, 10, 15, 21 © Royalty-Free/Digital Stock; pp. (graphics), 6, 9, 11, 12, 14, 16, 18, 20, 23, 25, 27, 28, 31, 33, 34, 35, 37, 39, 41, 42; p. 13 (top) © Michael & Patricia Fogden/Corbis; p. 13 (bottom) © Gallo Images/Corbis; p. 26 © Bettmann/Corbis; p. 32 © Herb Segars/Animals Animals; p. 35 © Joe McDonald/Corbis

Designer: Nelson Sá; Editor: Wayne Anderson
Photo Researcher: Nelson Sá